DATE DUE

DEC 3 0 1991	AUG 1 7 2001	
APR 1 1992	AUG 1 7 2005	
MAY 2 7 1992		
JUL 3 1 1992		
JAN 3 1 1993		
APR 2 9 1993		
JUN 1 1994		
JUL 7 1994		
MAY 3 1 1995		
DEC 29 1995		
AUG 31 1996		
MAY 3 0 1997		
MAY 0 7 1998		
MAR 0 7 2000		
MAY 3 0 2000		

HIGHSMITH #45220

WITHDRAWN

S0-EAY-695

© 1990 Franklin Watts

Franklin Watts Inc.
387 Park Avenue South
New York, NY 10016

Editor: Hazel Poole
Design: K and Co
Consultant: Michael Chinery

Printed in Belgium
All rights reserved

Library of Congress Cataloging-in-Publication Data

Watts, Barrie.
  24 hours on a seashore/Barrie Watts.
     p.   cm. — (24 hours)
  Summary: Relates what happens to the plants and animals along the seashore during a twenty-four hour period.
  ISBN 0-531-14037-7
  1. Seashore biology — Juvenile literature. [1. Seashore biology.]
I. Title. II. Title: Twenty-four hours on a seashore. III. Series.
QH95.7.W37   1990
574.5'2638 — dc20   89-38998
                                                                                                 CIP
                                                                                                  AC

24 HOURS ON A SEASHORE

Text and photography by Barrie Watts

FRANKLIN WATTS
NEW YORK · LONDON · SYDNEY · TORONTO

CONTENTS

🌙 **Early Morning**   **7**

☀️ **Daytime**   **15**

🌘 **Evening**   **27**

🌑 **Night**   **35**

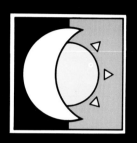

Early Morning

As the sun rises,
life on the seashore begins to stir ...

Early Morning

It is early morning on the seashore and the start of a new day. The tide is beginning to go out and many of the creatures that live on the shore are searching for shelter from the rising sun.

Few seashore animals can tolerate the direct heat of the sun. Most of them need moist, shady conditions if they are to survive when the tide goes out. The hard rock and stones of the seashore means that they are not able to burrow like the inhabitants of sandy beaches. Instead they have to stay where they are or hide themselves away so they do not dry out. Crabs and little fish called blennies hide in tiny cracks and crevices in the rocks, and, as long as their hide-outs remain damp, the animals can survive the few hours until the tide comes back in.

▼ Starfish are easily stranded by the outgoing tide because they are unable to move quickly. They will survive as long as they are not dried by the sun.

◀ The Devil or Velvet Swimming Crab waves its pincers in the air as a defense when disturbed. (Below left) Common shore crabs hide under rocks and seaweed at low tide.

As the tide goes out, it leaves small pools of seawater that have been trapped by rocks and stones. These natural rockpools provide homes and shelter for some of the animals that live on the upper and middle parts of the seashore. Many of them will live near the food they eat. To move unnecessarily, whether covered by the sea or not is dangerous. Predators are more likely to see their prey if it moves. The animals that hide themselves stand a better chance of not being eaten.

▲ Blennies are able to survive for long periods out of water as long as they are sheltered in a damp crevice or under seaweed. This one will spend at least five hours out of water and is kept alive by a trickle of seawater draining from a rockpool.

Early Morning

The receding tide leaves a line of seaweed on the shore, together with seashells and the remains of crabs and other dead creatures. Gulls and wading birds like the oystercatcher find plenty to eat here. They will spend all day searching for tasty morsels along the tideline. Clusters of mussels and other shellfish can be seen on the rocks as the tide goes out.

In a quiet secluded bay, gray seals have come ashore to give birth and rear their young. At birth, the babies weigh only about 14 kg (31 lb). After 16 to 21 days of feeding on their mother's rich milk, they will have grown to about 43 kg (95 lb). They are then abandoned by the mother and as they get hungry, they enter the sea to search for food.

▶ Gray seals will come ashore to give birth only on remote and quiet parts of the seashore. This youngster is about two months old and has already started to catch its own food.

▼ The oystercatcher searches for food along the tideline. It also eats cockles and mussels but, despite its name, it does not eat oysters. (Inset) Mussels are bivalves and live in tightly packed groups. They will open up their shells when covered by the sea so that they can feed.

Early Morning

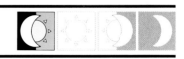

Many creatures retreat into their shells as soon as they are uncovered by the sea.

The limpet returns to exactly the same spot on its rock every time the tide goes out. It clamps itself down with its large, muscular foot. Barnacles look lifeless when they close their shells as the tide goes out, but the animals inside are quite safe because no water can escape from their almost air-tight shells. They open again when the tide returns and the animals strain tiny food particles from the water.

Dog Whelks need more shelter so they will seek out the damper, darker places and remain there until the tide returns.

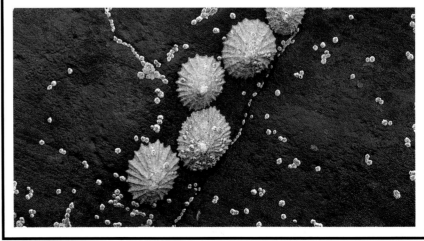

▲ Dog Whelks eat barnacles and mussels so are almost always found in the same places on the seashore. Their flask-shaped egg capsules can contain up to one thousand eggs. Out of these only about a dozen will survive.
(Top right) There can be as many as 100,000 barnacles per square mile on some rocky seashores.

▶ Limpets grind their shells to fit the rocks on which they live. This tight fit enables them to retain seawater at low tide.

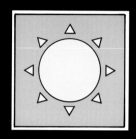

Daytime

With the rising temperature,
the shoreline appears almost deserted...

Daytime

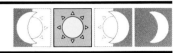

As the sun rises, the warm sunlight penetrates many of the rockpools on the shore. They quickly heat up, and soon the water temperature in some will have risen by up to 10°C. This reduces the oxygen in the water unless there is seaweed growing in the pools, as seaweed gives off oxygen. Only a few special creatures can tolerate these rapidly changing conditions. Shrimp, small fish like gobies and certain sea anemones are able to live in these pools on the higher part of the seashore.

Each seashore will have its own selection of creatures living on it, because each species of animal prefers to live under certain conditions. Wave action, exposure to sunlight and wind, seaweed growth and types of rock all dictate what lives on a seashore.

▼ This rockpool is in the higher zone on the seashore. Shrimp, gobies and snakelocks anemones can live in this type of rockpool, even though it heats up rapidly in the sun and cools rapidly at night when the tide comes back in.

The more varied the geography of a seashore is, the greater variety of animal life there will be living on it. As each type of creature likes to live under certain conditions, the seashore can be split into three zones, higher, middle and lower. Small periwinkles and some barnacles live above the higher zone. They are only covered by the sea on very high spring tides.

Common limpets can be found in all zones, while sea urchins, dahlia anemones and rock crabs are normally found in low tide zones.

Creatures found in the higher and middle zones spend more time out of water than those in the lower and need to ensure that they do not dry out.

▲ Coral algae is a brittle type of seaweed living in shallow, sunny rockpools. Snakelocks sea anemones like the same conditions so they are often found living together.

(Inset) Shore crabs are often found hiding under seaweed and rocks in rockpools in all three zones of the seashore.

20 Daytime

▶ Beadlet anemones are common on the seashore and live in all the zones. They are able to retract their tentacles out of water.

▼ Hermit crabs do not have shells of their own so they use empty periwinkle shells when they are young and move to larger whelk shells as they get bigger.

At midday the sea has reached its lowest level for two weeks. The tides that ebb and flow on the seashore today are called spring tides. These occur every two weeks throughout the year and are the best time to explore the seashore. Rockpools which are not normally uncovered by the sea can now be studied. On the bottom of these lower pools are strange colorful crusts made by certain red seaweeds. If they are exposed to the sun, however, they die and eventually turn white.

▼ Rockpools in the lower zone normally have much more seaweed growing in them, so they are much darker. Rockweed covers the bottom of these pools when the tide is out.

Daytime

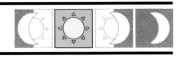

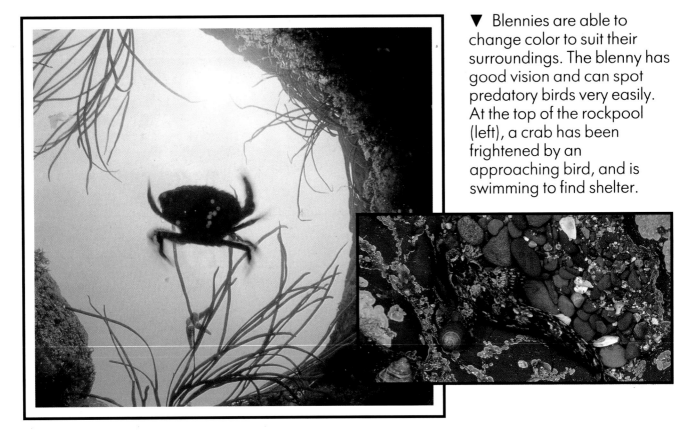

▼ Blennies are able to change color to suit their surroundings. The blenny has good vision and can spot predatory birds very easily. At the top of the rockpool (left), a crab has been frightened by an approaching bird, and is swimming to find shelter.

The rockpools appear deserted during the daytime, but many creatures are hiding beneath the rocks and seaweeds. Shrimp are almost transparent, so they are difficult to see unless they move. Blennies have no body scales like most other fish and are able to change their color to blend in with their surroundings.

Camouflage is very important, as the biggest threat to the smaller animals comes from the larger ones.

Even when the tide is in during the day, the underwater life of the seashore is wary of moving. The safer cover of the dark night will be the time of maximum activity. Then, the seashore will come alive. Crabs, shrimp and small fish will come out of hiding and move about in relative safety. They will feed on mollusks, seaweed and sometimes on each other.

▶ Starfish are some of the most colorful animals to be found on the seashore. The common starfish is about 10 cm (4 inches) across and is found hidden among seaweeds in the lower zones. The cushion starlet is only 2 cm (³/₄ inch) across and is found hidden under rocks and stones all down the seashore.

Daytime

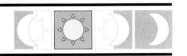

Throughout the year the seashore is a natural food store for many different kinds of animal. Birds, in particular, use the seashore as a place to breed, rest and feed. Some, like the herring gull and oystercatcher will be present all year around. For wading birds like the sandpiper, the shore is a vital winter habitat. Great flocks of waders congregate on shingle banks and seashores at high tide. They sleep until the tide goes out and they can feed again.

Seashore plants provide food all year around for insects and birds. Butterflies and beetles are attracted to large, flowering clumps of sea campion, thrift and thistles during early summer. Later in the year, the seeds of these plants provide food for several species of seed-eating birds such as the goldfinch and the seaside sparrow.

▶ Many birds such as the sandpiper will use the seashore as a temporary resting and feeding place as they fly south for the winter.

▼ The seaside sparrow visits the seashore to feed on the seeds of plants growing on sand and shingle at the top of the shore.

26 Daytime

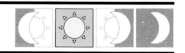

At two o'clock the sea is at its lowest on the seashore. Oystercatchers are working along the tide level searching and probing for cockles and other mollusks washed up by the tide.

On the sandy beach, dead sea animals have been stranded by the tide. At certain times of the year when the sea is rough, jellyfish are left on the seashore. They are slow swimmers and therefore are easily swept along by the ocean's strong currents.

Old crab shells and seashells are also left behind on the shore and will be broken up by the movement of the waves and the grinding sand. The tiny pieces will then form part of the seashore where they had originally started life.

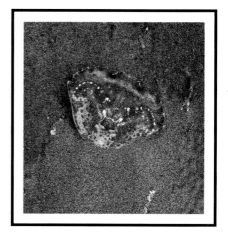

▲ A crab shell is called a carapace. When a crab dies, its carapace is often washed onto the seashore. It is soon broken into small pieces by the crashing waves.

◀ Jellyfish can grow to more than 40 cm (16 inches) across. Most of them feed on the plankton that drift with them in the upper layers of the sea.

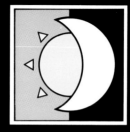

Evening

It is evening,
the tide is rising,
and some of the rockpool creatures start to em

30 Evening

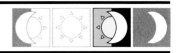

Early evening on the seashore and the tide is already well on its way back up the shore. As the bright daylight begins to fade, some of the rockpool creatures emerge from hiding.

The pools on the lower shore have now had new seawater washed into them by the incoming waves. This has brought a fresh supply of food and oxygen to them, so anemones, shrimp and fish begin to feed after several hours of inactivity.

Starfish are able to sense that night is approaching because the tips of their arms are light sensitive. They, too, begin to emerge from their hiding places to look for food.

▶ The beadlet anemone has a large flexible mouth which it uses to swallow prey caught by its stinging tentacles. It has a large sucker which enables it to cling to the rocks.

▼ The sea scorpion is well camouflaged and food is easily caught in its large mouth. Its spiny fins also protect it from being eaten by any predator.

▲ The shrimp is hard to spot unless it moves. It will emerge to feed in the evening although it will not feel safe until the night comes.

◄ The blenny starts to feed as soon as the tide washes it from its hiding place where it remained during the day. Small crabs and mollusks are snapped up, but its main food is barnacles.

Blennies swim out of their crevices as the sea rises over them. Their favorite food is barnacles, which they eat by biting them off the rocks with their sharp teeth.

The bully of the rockpools is the scorpion fish. It is a slow swimmer but is a fearsome predator of small blennies, gobies and shrimp. It simply waits for its prey to swim close by and then it gulps them into its large cavernous mouth.

32 Evening

◀ Gray plovers stop on the seashore to rest as they fly south for the winter. Their hunched posture and sad expression make them look unhappy.

▼ The rising tide has forced these oystercatchers to stop feeding and to group together. They will often feed at night when the tide is low.

As the tide rises, birds are forced higher and higher up the seashore until they abandon their feeding. Some birds, however, such as the ringed plover, continue to feed along the moving tideline. They will eventually roost with the other birds on the shore until they can return to their feeding areas.

Large numbers of oystercatchers, sandpiper and plovers group together at roosting time. They usually use the same spot on the seashore each day to rest and sleep.

Sometimes they sleep with their head or bill tucked under one wing, their eyes constantly opening and closing, ever watchful for a swooping bird of prey or annoying crow, as danger is never very far away.

Evening

When the sleeping birds are disturbed they will take flight instantly. Their crying alarm calls will alert other roosting birds along the seashore. They will spend most of their roosting period flying from one place to another if danger threatens. If this happens, some of the smaller birds like the sandpiper could get exhausted and die. They need plenty of food and rest each day when they are migrating.

Some of them will feed at night when the tide is low, and will use their sensitive long bills to probe for worms, mussels and small crabs.

▼ This sandpiper is sleeping. It will keep opening its eyes and does not often sleep soundly. If one of the roosting birds becomes frightened, all the rest will fly off as well.

Night

Under the cover of darkness,
the rockpools come to life...

38 Night

Night has come to the seashore and the tide is already on its way out again. The seashore animals feel safe now, because most of the predatory birds are sleeping.

Crabs crawl out from their hiding places to scavenge on dead creatures washed in by the last high tide. They will also catch and eat worms and other slow-moving animals.

Limpets crawl over the rocks to browse on the algae even after the tide has uncovered them. There is also plenty of activity among the inhabitants of the rockpools.

▼ This common sea urchin is crawling over a patch of mussels. Its tube feet enable it to move about. The urchin's mouth is on the flat underside of its body and it uses a ring of sharp, beak-like teeth to scrape algae from rocks and shells.

▲ This hermit crab is fully grown and living in a large whelk shell. The shell has barnacles and other small creatures growing on it.

(Top right) The starfish uses tube feet to move about, just like the sea urchin. It feeds on mussels and cockles.

▶ Young rock crabs spend most of their life on the lower shore. As they get bigger they seek deeper water. Like other crabs, should it lose a leg it will simply grow another one.

Night

Throughout the night the rockpools are a hive of activity, whether the tide is in or out. Sea gooseberries are often stranded in them and will catch plankton with their two retractable sticky tentacles.

Barnacles are also busy feeding while the tide still covers them. They use their feathery feet, called cirri, to sieve plankton from the flowing seawater.

Green sea lettuce is a favorite food of the common shrimp. Shrimp have two pairs of sensitive antennae which are longer than their body and enable them to feel for food in front and behind them at the same time. It uses its pair of pincers to tear its food into bite-sized pieces. When disturbed, it shoots backwards rapidly helped by its large fanned tail and paddle-like legs.

(Below left) The sea gooseberry is as big as a grape and feeds on small crustaceans that live in the sea's plankton layers. It moves its jelly-like body through the water by using rows of tiny paddles on its sides.

(Below right) The barnacle feeds by using feathery feet which move in and out rapidly over 100 times a minute.

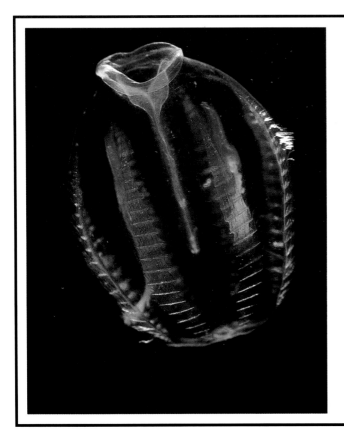

◀ Shrimp usually only feed at night. Their beady eyes are on stalks which make it easy for them to spot predators.

▼ Mussels feed by pumping up to 10 gallons of seawater each day through their frilly gills and straining plankton from it.

Many creatures on the seashore rely on wave action and the twice daily high tide to bring food to them. Microscopic plankton is carried in the upper layers of the oceans. It is the main food for many animals on the seashore. Mussels sieve and filter seawater with their frilly gills to extract these fine particles of food.

Plankton grows in the sea and is nourished by the chemicals found in seawater. If the sea is polluted, this could have disastrous effects. The plankton could easily die, the food chain will collapse and the mollusks and birds which live on the seashore will eventually disappear.

Night

No two days on the seashore will be alike. The tides rise and fall at different times each day. During the winter, the cold weather will make many animals seek deeper waters because they cannot survive the cold.

Summer and spring are the best times to visit the seashore because most creatures will seek the warmer, shallower water to lay their eggs and breed. The variety of life present during this time also encourages other animals to visit and feed. All are closely dependent upon each other and they rely on the sea to be clean and free from pollution to sustain them.

▼ The sea lemon, which is actually a kind of sea slug, hides beneath rocks and in crevices when the tide is out. When covered by the sea at night, it emerges to feed on sponges on the seashore. Its beautiful feathery gills are on its back. It glides gracefully over seaweed in its search for food.

Glossary

Antennae
Feelers or horns on insects and crustaceans.

Bivalve
A mollusk whose soft body is held within a pair of similar hinged shells. Examples: clams, mussels, oysters.

Camouflage
The color and shape of an animal's shell or skin that enable it to blend with its habitat to hide from predators.

Capsule
Small container.

Carapace
Shell of a crustacean.

Cavernous
Deep and hollow.

Cirri
Feathery feet of barnacle used for catching food.

Crevice
A crack or narrow opening.

Crustacean
Animal living in water that has a hard shell and a body divided into lots of segments. Examples: crabs, shrimp, lobsters.

Feeding Areas
Areas of land where animals and birds regularly return to search for food.

Food chain
A chain of animals feeding on each other. Each depends on a part of the food chain to survive. Should one part of the chain die, the whole chain will collapse.

Habitat
The normal home or locality of animals and plants.

Microscopic
Can only be seen with a microscope.

Migrating
Flying to a different country and climate.

Mollusk
A soft-bodied animal that normally has a hard outer shell. Examples: clams, oysters, mussels, snails, slugs.

Morsels
Small pieces of food.

Pincers
A limb or arm with a pair of claws to grip objects.

Plankton
Microscopic animals and plants that drift together in upper layers of oceans.

Predator
An animal that eats or preys on another.

Recede
To draw back.

Retractable
Able to be drawn back in.

Scavenger
An animal feeding on the dead remains of other animals.

Shingle
Small, weather-worn pebbles found especially on beaches.

Spring Tide
Tides that occur every two weeks when the earth, moon and sun are in a straight line. The sea will move very high up and very low down the seashore at each spring tide.

Tentacles
Long, flexible arms that feel and grip food and prey.

Tube feet
Long, flexible suckers that enable sand dollars, sea urchins and starfish to move along the seashore.

Wave Action
The movement of the tide.

Zone
Part or area of the seashore.

Index

anemones 30

barnacles 14, 31, 39, 40
beadlet anemone 20, 30, 31
birds 12, 24, 25, 26, 32, 33, 34, 43
bivalve 43
blennies 11, 22, 30, 31

camouflage 23, 43
capsule 14, 43
carapace 26, 43
cirri 40, 43
common shore crab 11, 19
coral algae 19
crabs 11, 22, 26, 38, 39
crustacean 40, 43
cushion starlet 23

devil crab 11
dog whelk 14

eggs 14, 42

fish 30
food chain 41, 43

gobies 18, 30
goldfinch 24
green sea lettuce 40
gray seal 12, 13
gulls 12

hermit crab 20, 39

jellyfish 26

limpet 14

mollusk 43
mussels 12, 26, 38, 39, 41

oyster catcher 12, 26, 32, 33

periwinkles 19, 20
pincers 11, 43
plankton 26, 40, 41, 43
plovers 32

rock crab 19, 39
rockpool 11, 18, 21, 22, 30, 38
rockweed 21

sandpiper 24, 25, 32, 34
scorpion fish 30
sea anemone 18
sea campion 24
sea gooseberry 40
sea lemon 42
sea urchin 38, 39
seabeet 25
seaside sparrow 24
seaweed 18
seed eating birds 24
shells 20, 26, 27, 39
shelter 10
shrimp 18, 22, 30, 31, 40, 41
snakelocks anemone 18, 19
sponges 42
starfish 10, 23, 39

thistles 25
tides 20, 30, 32, 34, 38, 40, 42, 43

PRINTED IN BELGIUM BY
proost
INTERNATIONAL BOOK PRODUCTION